BRAINSTORMING ON DESIGNS AND SCHEMES II

样板房设计新法（上）
NEW DESIGN METHODS OF SHOW FLAT

深圳视界文化传播有限公司 编
Shenzhen Design Vision Cultural Dissemination Co., Ltd

中国林业出版社
China Forestry Publishing House

PREFACE
序言

The house itself has no attributes. What make the space shine are always the residents who live in the space and the times that flow slowly in the space and stay in the hearts forever. Stories condensed in this exclusive space carry many dreams and hopes. For a designer, how to interpret the legend of the space by colors, structure designs, lights, shapes and different elements to make the residents feel cozy and comfortable so as to have material and spiritual attachments is a complicated question.

Home is a place with emotions and stories. There is a landmark with strong historic cultures and humanity deposits as if it is a land of idyllic beauty, where once remained many stories about princes, dukes and poets. Under the green shadow, there are unknown prosperity and luxury, and even the grace of a startled swan and a swimming dragon. In consideration of identity, the designer must try his best to choose and collocate the interior furnishings to design a house which confirms to the owner's identity. The house is carried forward like a towering tree and becomes eternal. It seems that people can see the traces of fleeting time and exquisite art deposits through the house.

The power of science and technology encourages the spread of futurism. People are surrounded by the living era. The designers integrate paintings, poems and dramas into the design and apply them to the limit. The design pattern of combining human factors and natural conditions with the modeling gradually becomes a representative form of design.

The design of show flat always stands ahead of interior decoration and design, and leads the trend of interior decoration and design. New design concepts, new materials, new design elements, new crafts and new displays are presented in the design of show flat. The popular styles, environmental protection concepts, energy saving materials and displaying collocations of show flat design are always the wind vane to lead consumers.

The original spirit of show flat design is very important. The designers concentrate on the distinct theme to bring the new concepts and ideal living atmosphere to the residents so as to promote people's living quality. The designers want to express their unique views on design, try to perform personal and unique charm, and reflect the distinct personality of the building in the show flat design.

Shanghai YST Design Co., Ltd
Vivian

EMOTIONAL ART SPACE
情感的艺术空间

房子本身是没有属性的，让一个 space 熠熠发光的永远是居于其中的人，还有在这个空间下缓缓流淌却永存心间的时光。凝结于这个专属空间中的故事，承载着太多的梦想和期望。在设计师的笔触下，如何将这个空间的传奇以色彩、结构设计、灯光、形状以及不同的元素来演绎，让客户在整体感觉上安逸舒适，产生物质上和精神上的眷恋感，是一道复杂的简答题。

家是一个有情感有故事的地方，有一处地标拥有着浓厚的历史文化和人文沉淀，像一方世外桃源，那里曾经留下了许多关于王子、公爵、诗人的故事。在绿荫的掩映下，隐藏着不为人知的繁荣与奢侈，更有翩若惊鸿，婉若游龙的气质。因为身份这一层关系，设计师必须在室内装饰的选用搭配上，发挥自己最大的功力，设计出与主人身份相符的房子。使房子像参天大树一样繁茂传承下去，成为永恒，而通过房子人们仿佛可以看到岁月流过的痕迹和绝美的艺术沉积。

科技的力量助长了未来主义风潮，人们被生活时代感包围着，设计师们在创作时把绘画、诗歌、戏剧都融合到设计中，并且把它们发挥到极致，将人文因素和自然条件与构思造型相结合，这种设计形式逐渐发展成具有代表性的设计形式。

样板间设计往往是站在室内空间装饰、装修设计的前沿，引导着室内空间装饰与装修设计的潮流。新的设计理念、新的材料、新的设计元素、新的工艺、新的陈设在样板间设计中都会体现。样板间设计的流行风格、环保、节能材料、陈设搭配，往往是引领消费者的风向标。

样板间设计，原创精神是很重要的。设计师要紧紧围绕鲜明的主题，把新的理念、理想的人居环境带给置业者，这样才能促进人们生活品质的不断提高。设计师要表达出自己对设计特有的看法，努力表现出个性化、唯我的独特魅力，并在样板间设计中反映出楼盘鲜明的自我个性。

上海益善堂装饰设计有限公司
宋莹

CONTENTS
目录

BRITISH & FRENCH STYLE
英法风格

008
REPUTATION KEEPS SHINNING IN THE YEARS
岁月流芳

018
NEW PHILOSOPHY OF BEAUTIFUL HOME
新派美居哲学

024
THE ENCOUNTER BETWEEN LIFE AND ART
当生活与艺术邂逅

034
LAKESIDE GARDEN
花园湖畔

046
THE HARBOR OF HEART
心灵的港湾

052
THE SMELL OF PEPPERMINT
薄荷的气息

MODERN LUXURIOUS STYLE
现代奢华风格

066
THE CHARM OF SIMPLE LUXURY
简奢的魅力

074
FLEXIBLE NORDICISM
柔性北欧主义

086
THE THRIVING DREAM
茁·梦

094
ART SPACE
艺术空间

104
IMPRESSIVE SYMPHONY
印象交响乐

116
THE INTERWEAVE OF LUXURIOUS ELEGANCE AND MODERN SIMPLICITY
奢华典雅与现代简约交织

124
ART PARTY
艺术派对

132
QUALITY LIVING
品质生活

140
NIFTY MAGICLISM
俏皮魔幻主义

158
PAINTING THE PHANTOM OF CITY
绮绘都市魅影

SIMPLE EUROPEAN STYLE
简欧风格

166
NEW INTERPRETATION OF FASHION
时尚新解读

176
THE JUMPING COLORS
跳跃的色彩

186
TONE
格调

198
EUROPEAN STYLE
欧式风情

206
A RESIDENCE WITH SUNNY VIEW
向阳而居

220
SUNNY RESIDENCE
阳光住宅

228
THE COLLISION OF ART
艺术的碰撞

MODERN CHINESE STYLE
现代中式风格

236
THE ANNUAL RING
年轮

244
MODERN NEW HOME
时代新居

252
LOOKING FOR ORIENTAL FLAVOR : A SECRET FRAGRANCE
暗香·寻觅东方韵

268
ATTACHMENTS OF BEAUTIFUL SCENERIES, POSSESSIONS OF TIMES
眷红偎翠，私藏岁月

276
ORIENTAL CHARM
东方韵味

284
A CHARMING NEW HOME IN SUZHOU
苏韵新居

294
THE MEMORY OF PATIO
天井的记忆

MEDITERRANEAN STYLE
地中海风格

302
SKY AND SEA
海天一色

310
FRESH BREEZES BLOW GENTLY
清风徐来

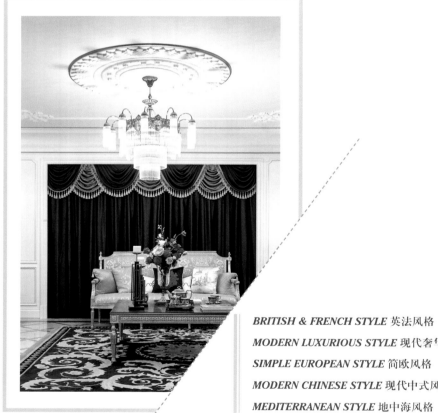

BRITISH & FRENCH STYLE 英法风格
MODERN LUXURIOUS STYLE 现代奢华风格
SIMPLE EUROPEAN STYLE 简欧风格
MODERN CHINESE STYLE 现代中式风格
MEDITERRANEAN STYLE 地中海风格

BRITISH & FRENCH STYLE
英法风格

REPUTATION KEEPS SHINNING IN THE YEARS
岁月流芳

设计公司：上海元柏建筑设计事务所
设 计 师：史迪威
项目面积：778平方米
项目地点：上海
主要材料：装饰画、大理石、木地板等

DESIGN CONCEPT / 设计理念

>> For historical architectures or classic artworks, what make them immortal are the deep aesthetics which can overcome the challenges of fashion, and high-grade temperament. Home is a place where actual scenes of life take place, but with the designer's distinctive expression of aesthetics couples with flexible use of materials and perfect visual proportions, a house beyond low-key luxuriousness is displayed wonderfully. The design of the whole space reveals a high-grade smell of low-key luxuriousness, it is themed with light colors. The design methods start from the basic open space and the conformability of ergonomics, every procedure shows the elegant atmosphere, with various environmental materials, it manifests high-grade and humane coziness.

>　　无论是历史悠久的传世建筑还是经典的艺术作品，经得起潮流考验的深度美感与内在的高品位气质，才是它们跨越时间得以长存的两大特征。家虽然是人们实际居住的生活场景，但透过设计者对于美感的独特表述，加上灵活的材质运用和完美的视觉比例适当地辅助，精彩呈现出低调奢华之外的名家向度。整个空间的设计都透露出一种低调奢华的高品位气息，设计中以浅色为主旋律展开联想。设计手法从最基本的空间开放感觉与人体工程学的舒适角度出发，每个环节都透露出高雅的气质，加以各类环保材料融合在一起，流露出高品位的人性化舒适感。

NEW PHILOSOPHY OF BEAUTIFUL HOME

新派美居哲学

设计公司：深圳臻品设计顾问有限公司
陈设软装：深圳臻品设计顾问有限公司
项目面积：149平方米
项目地点：广东东莞
主要材料：大理石、地毯、木地板等

DESIGN CONCEPT / 设计理念

>> The design of this project is warm and exquisite, skillfully integrates the comfort of big space with the charming of little fun and conveys the pursuit of a better life and the comfort of furnishing art, making the designs full of material needs as well as spiritual pursuit. Let's enjoy the high pace life, experience the elegance and taste the temperament. According to elegant colors and languages, it is based on classical elegance, embeds new European aristocratic style and is integrated into an exquisite Western hotel-style living space.

>> 本案整体设计温暖精致，把大空间的舒适和小情调的魅力巧妙地融合，把对美好生活的追求和陈设艺术的写意通过点滴精妙呈现，使空间的设计既满足物质需求更追求精神层次！让我们在享受生活快节奏的同时，也体会一份优雅，品位一种气质。循着优雅的设色及语汇，以古典主义的优雅细致为线，镶入欧式新贵族风范，糅合成宛如西方精品酒店般的居住空间。

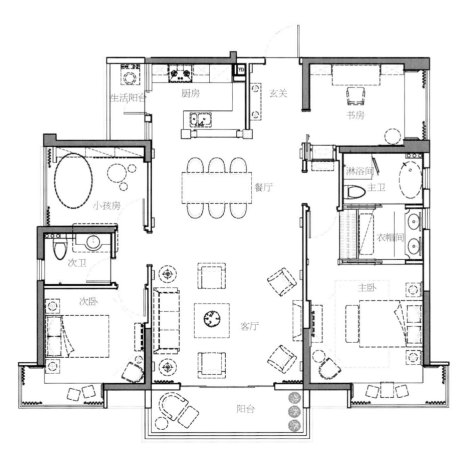

The graceful lights in the living room pour down; the classical modeling becomes an indispensable art in the space. The green curtains dance with the wind, just like flowing colors, decorating every detail of life. There are some bright orange and green in the space, presenting the romance and comfort of the new aristocratic life.

客厅中央曼妙的灯光随影而下，带有古典韵味的造型，演化成了空间中不可或缺的艺术品。绿色的窗帘在微风的荡漾下飘扬，宛如流动的色彩，装点生活的每个细节。空间里不乏亮丽的橘色与绿色点缀，处处点精，展露出浪漫安逸的新贵族生活。

THE ENCOUNTER BETWEEN LIFE AND ART
当生活与艺术邂逅

设计公司：Pin-Design 致品空间
设 计 师：Mary Cai（蔡智萍）、Pin-Design 设计团队
项目面积：260 平方米
项目地点：江苏太仓
主要材料：法国米黄大理石、霸王花大理石柱、大西洋灰大理石、老矿金花大理石、橄榄灰大理石、白玉兰大理石等

DESIGN CONCEPT / 设计理念

>> *Variations on a Rococo Theme* created by Tchaikovsky is one of the most classic works in the history of classical music. In the fluent, gorgeous and cozy melody, there seems to be a graceful figure of a noble beauty who went through many things. Inspired by this classic composition, the designer integrates the dynamic and emotional music with the essence of classical Rocco art by elegant techniques. Here comes a gorgeous chapter of encounter between life and art in the gradually changed rhythm.

>> 由柴可夫斯基创作的《洛可可主题变奏曲》是古典音乐史上最经典的作品之一，流丽安逸的旋律中依稀可见一位贵族丽人历经世事而优雅醇熟的瑰丽身影。设计师以此经典乐曲为灵感，以优雅手法将音乐的动感与情感、洛可可古典艺术精粹在空间中融合，于渐次生变的韵律中谱写出一曲生活与艺术邂逅的华彩篇章。

The designer regards new humanism aesthetics as spirits of artistic creation and integrates modern languages with classical essences to create the exclusive aesthetics and delicate attitudes of the space. After the modification of the original house type, the interiors follow the classical structurology and apply harmonious symmetrical layout and rigorous classical pillar-shaped structure, displaying the honor, generosity and elegance of the big flat completely. Two sides of the corridor are open living room and dining room respectively. The towering marble Corinthian pillars make the space noble and grand. Exquisite gilded chapters with rolling floral leaf patterns diffuse calm and noble temperament like artworks.

设计师以新人文主义美学为艺术创作精神，将现代语汇与古典精粹巧妙融汇，以此表达空间自身独到的审美与精致态度。在对原有房型进行改造后，室内遵循古典主义结构学，采用和谐的对称布局和严谨的古典柱式构图，将大平层住宅的尊贵、大气、典雅表现得一览无余。走廊两侧是相对的开放式客厅与餐厅，高耸的大理石科林斯柱将高贵恢弘的气势注入空间，精美的描金卷涡花叶纹柱头，宛如艺术品般散发沉静高贵的气质。

LAKESIDE GARDEN
花园湖畔

项目名称：杭州绿城曼陀花园
设计公司：尚层装饰（北京）有限公司杭州分公司
设 计 师：王贺麟
项目面积：350 平方米
项目地点：浙江杭州
主要材料：米黄系大理石、象牙白混油饰面、美国红樱桃木饰面、米黄系壁纸饰面等

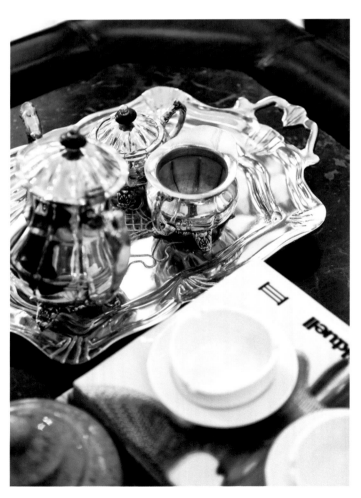

DESIGN CONCEPT / 设计理念

>> This residence locates near the Qingshan Lake in Linan, Hangzhou, where the geographic and natural environment is richly endowed by nature. Though it is a show flat focusing on presenting high-quality living atmosphere, it decreases some complicated and fancy interior designs based on the French style, remains the classic French elements and symbols to get close to real life, and creates a relaxing and comfortable feeling. The finalized solution uses newly clarified axis relationship to add classic design symbols to strengthen the boundary of the space in the crucial separating point.

>> 本案位于杭州临安青山湖畔，地理自然环境得天独厚。项目营造虽说是样板房，以展示高品质的生活氛围为基调，但在室内法式风格为主的前提下尽可能地去掉一些繁琐和花俏的设计，保留经典的法式元素符号，让整体的设计氛围更贴近我们的实际生活，营造出"放松、舒适"的感觉。定稿方案利用重新梳理的轴线关系，在关键的分界处加入以强化空间界线的经典的设计符号。

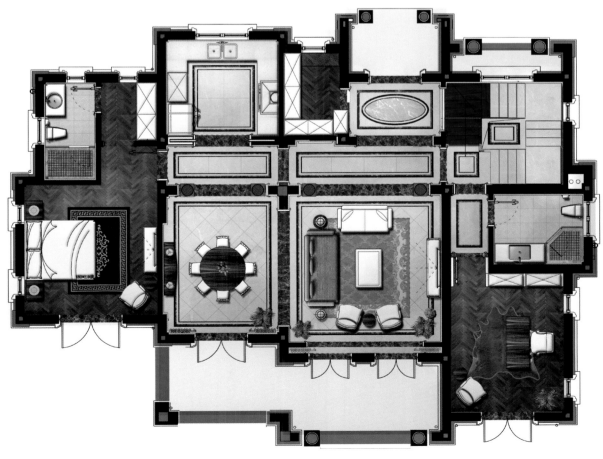

In the phase of floor planning, the designer mainly solves the problems of space disorders and space functions which don't meet the needs of real life. The designer focuses on adjusting the functions of the first floor. The bathroom at the entry door is converted into a storeroom to cover the shortage of placing clothes and hats. One elder's room en suite is remained, while the other is converted into a study and a public bathroom. The adjustments enrich practical functions of the space and add an ornamental value as the main spaces of a villa. At first the relationship of opposite scenery between the interior and outdoor is not ideal and is redesigned to keep the harmony and smoothness of the interior and outdoor.

我们在平面规划阶段，主要解决空间凌乱且功能设置不符合实际生活用途等问题。我们重点把一层的功能进行了调整。入户门的部分将一个卫生间改成了储藏室，以弥补入户鞋帽储藏量的不足。保留了一个老人房套间，另外的一个套间调整为书房和公共卫生间。这样既丰富了空间功能使其更贴近生活又增加了作为别墅的主要空间的观赏性。另外室内外的对景关系都不够理想，所以重新进行了梳理，以保障室内外气场的和谐畅通性。

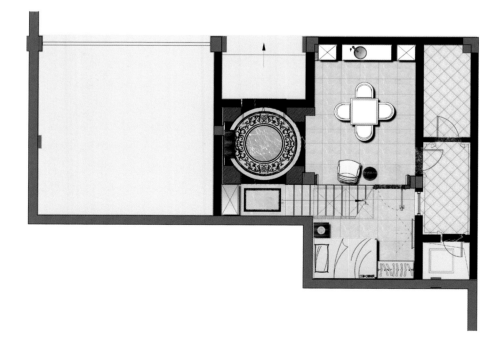

THE HARBOR OF HEART

心灵的港湾

设计公司：乐摩装饰设计
设计总监：潘江
主设计师：张妍
项目面积：280平方米
项目地点：江苏苏州
主要材料：地毯、大理石、吊灯、护墙板等
摄 影 师：金霈

DESIGN CONCEPT / 设计理念

>> Home is the harbor of heart. In this city with reinforced concrete, we are longing for a comfortable home with large French windows where we could enjoy leisure time with family members when reclining on the sofa. On the entire space planning, the designers take the big unit into consideration, decorating the house in a compact and full pattern, so that the overall coordination of furniture and space is maximized. The soft blue color is applied throughout the entire space, which is peace and cozy, making the home full of elegance and happiness.

>> 家，是心灵的港湾。在这个钢筋混凝土的城市，我们渴望有一个舒适自在的家，大的落地窗，慵懒的靠在沙发上与家人谈笑家常，乐享休闲时光。在整个空间布局上，充分考虑到大户型的特点，以紧凑而饱满的节奏进行装点，最大限度地体现家具与空间的整体协调性，以温婉柔和的蓝色为基调，贯穿整个空间，安静而惬意，让家中时刻洋溢着淡淡的优雅与幸福的味道。

The lusterless furniture corresponds the hard decoration lines, the application of bronze and pottery textures reveal the elegance and nobility of space naturally, providing the original warmth and safety of home. The orange color on the sofa looks like the warm light in the nightfall, tender and soft, guiding the way home. The lazy recliner and classical backboard show the unique temperament and taste of the master bedroom when matching with the mature and fascinating chocolate brown and light purple. The main bathroom is hidden in the master bedroom, a private space as it is, it has a beautiful lake view of Banyue Bay, which adds some romantic feeling. In the daughter's room, you may feel the soft and exquisite design in ever corner, the bright yellow enlightens the happy elements of the space, the blue-green keynote looks like the clean sky after the rain, which is azure, distant, pure...

哑光质感的家具对应硬装线条，古铜与陶瓷材质的引入，在收放自如中自然流露出空间的雅致与高贵，给予家最原始的温暖与安全。沙发上一抹橘色，如夜幕下温暖的灯火，不焦躁，不刺眼，指引着归家的路。慵懒的贵妃榻，古典的床背板，搭配成熟迷人的巧克力棕色与淡淡的紫，传递出主卧的独特气质及品位。而隐藏在主卧内，既能坐拥半月湾湖景，同时极私密的主卫，也为空间增添了一份小小的浪漫情怀。女儿房，温柔细腻的心思洒落在空间的每个角落，明快的黄色点燃空间的快乐因子，蓝绿色的基调犹如雨后一尘不染的天空，湛蓝、悠远、纯净……

THE SMELL OF PEPPERMINT

薄荷的气息

陈设设计：北京中合深美装饰工程设计有限公司
设 计 师：王卓娅
项目面积：190平方米
项目地点：山东新泰
主要材料：地毯、壁纸、大理石等

DESIGN CONCEPT // 设计理念

>> This project is based on elegant French style with retro and natural tone. The most prominent feature is the noble quality. The interior is classical and humanistic, whose inner temperaments are comfort, elegance and easiness. The main tone in the living room is fresh mint color, complemented with cream-color and embellished with faint yellow, which is free, tranquil, natural and gentle. The intellectual and sober gray, the moist flavor of grass and the bright and clean white make the whole space full of warmth and romance. The gentle and exquisite soft decorations extend to every corner of the space, which is elegant and comfort. Living here, one's heart can calm down. Tasting wine, brewing tea, reading books or taking a nap can be the great living atmosphere.

>> 本案以优雅法式风格定位，弥漫着复古、自然主义的格调，最突出的特征是贵族气十足。室内散发着古典和人文的气息，舒适、优雅、安逸是它的内在气质。客厅以清新的薄荷色为主色调，米色为辅助色，淡黄色为点缀色，自在宁静，自然柔和。三分灰的知性冷静，五分水润的青草气息，伴随两分白的皎洁明净，使整个室内充满温情与浪漫。软装温柔细腻的心思延伸到空间的每个角落，优雅而舒适，居住在这样的房间里，自己的内心也会沉静下来，无论是品酒、烹茶或者是看书、小憩，都是绝佳的居家情调。

061

BRITISH & FRENCH STYLE 英法风格
MODERN LUXURIOUS STYLE 现代奢华风格
SIMPLE EUROPEAN STYLE 简欧风格
MODERN CHINESE STYLE 现代中式风格
MEDITERRANEAN STYLE 地中海风格

MODERN LUXURIOUS STYLE
现代奢华风格

THE CHARM OF SIMPLE LUXURY
简奢的魅力

项目名称：深圳前海时代 F 户型样板房
设计公司：矩阵纵横
设 计 师：于鹏杰、王冠、刘建辉
项目面积：178 平方米
项目地点：广东深圳
主要材料：黑金花大理石、拿铁棕大理石、铜板、玛瑙石、黑檀烤漆木饰面、铜色屏风、银箔等

DESIGN CONCEPT / 设计理念

>> People feel the charm of simplicity in their journeys. When the exhausting body and mind are attached to home more intensely, what people want is a relaxing and free environment, focusing on the combination between different sizes of color lumps and blending regional decorations into the style. It uses various higher saturation of colors and many deep colors. Portopo, Latte coffee, agate, ebony and bronze are put together, presenting another color style. Art carpet, fur bedding articles, animal patterns and unique art flowers on the ground make the space extremely rich. The conduct of details applies classical Chinese elements, making the space more connotative.

>> 随着人们在旅游中感受到简约的魅力，当疲惫的身心对家的依恋越发强烈，人们想要的是轻松、自由的环境，注重大小色块间的组合，地域性的后期配饰融入设计风格之中。空间里大量使用了各类饱和度极高的颜色和大量深色，黑金花、拿铁棕、玛瑙、黑檀、古铜这些浓重的色彩放在一起却也迸发出另外一种色彩。艺术地毯、皮毛床上用品、动物花纹、独特的地面艺术拼花都使空间极其丰富。在一些细节的处理上也巧妙地运用了中国古典元素，让空间的文化底蕴十足。

073

FLEXIBLE NORDICISM
柔性北欧主义

设计公司：上海 ARCHI 意·嘉丰设计机构
设计团队：陈丹凌、彭兴、罗剑、何彩霞、陆艳、胡继伟
项目面积：260 平方米
项目地点：北京
主要材料：青玉大理石、雅兰大理石、橡木饰面板、拉丝黑金不锈钢、皮革、壁纸等
摄影公司：空间摄影工作室

DESIGN CONCEPT / 设计理念

>> Flexible Nordicism refines from the Nordic social design thoughts. Simple and humanized design languages, natural and clear materials and colors and the elegant flavors featured by the Scandinavia give people a feeling of home which is warm and comfortable. This project strengthens the warm impression by natural materials, such as wood, cotton and leather. It uses the earth tone which can give people a sense of security as the main tone, embellished with bright colors, which creates a comfortable and rigorous living atmosphere. At the same time, the designers customize various warm themes for different spaces to make everyone who comes in feel the strong loving atmosphere so as to trigger the longing and yearning to warm home.

>> 柔性北欧主义自北欧民主化的设计思想中提炼而出，简约且人性化的设计语汇，自然干净的材料、色彩，以及斯堪的纳维亚所特有的恬雅气息，给人以家的温馨与舒适感。本案通过木材、棉麻、皮革等天然材质强化其温情印象，并运用能予人安全感的大地色系作为主色调，以春天系亮彩色作为点缀，打造出温情脉脉又彰显活力的居住氛围。同时设计师还为不同空间定制了各种温馨主题，让每个走进来的人都能感受到浓浓的爱的氛围，并由此引发对温情之家的渴望与向往。

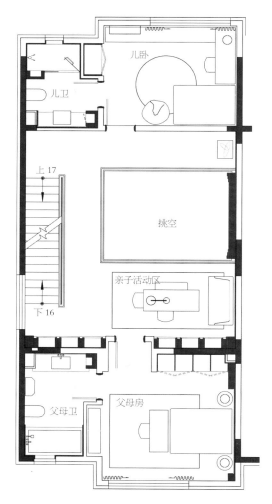

The walls of the interiors from the first floor to the third floor are mainly paved with wood, defining the natural and warm tone of the entire space. Contracted and plain furniture brings people back to the original state through simple modeling and color. Different spaces create funny living scenes by careful choice and collocation of furnishings. The living room is the place where the family collects travel memories and displays personal interest. The theme of dining room is the trip which the family is planning, which is relaxing and lively. The bright and open family activity room with geometrical elements is full of laughter from the family. The old-fashioned radio and camera in the elders' room engrave the fleeting time. The photo wall in the bedroom witnesses the family's precious memories for decades. Flexible Nordicism is a kind of design concept, which is carried by designs, materials and aesthetics, and finally completed by warm feelings.

室内从一层到三层大部分区域都采用木作墙面，界定出整个空间自然温暖的调性，简约中略带朴拙的家具，通过简单的造型与色彩带领人们回归本质状态。不同的空间则用精心挑选、搭配的陈设来营造充满趣味性的生活场景。客厅是一家人收藏旅行回忆和展示个人兴趣的地方；餐厅以正在策划的家庭旅行作为主题，轻松又活泼；以几何元素打造的色彩明快的开放式亲子活动区，回荡着家人们的欢声笑语；父母房里的老式收音机、相机等物品镌刻着流逝的悠悠时光；卧室里的照片墙则见证了一家人数十年来点滴的珍贵回忆。柔性北欧主义是一种设计理念，它通过设计、材料、美学来承载，但最终是靠温馨的情感来完成。

THE THRIVING DREAM

茁·梦

设计公司：深圳市盘石室内设计有限公司（吴文粒设计事务所）
陈设设计师：深圳市蒲草陈设艺术设计有限公司
主案设计师：吴文粒、陆伟英
参与设计师：陈东成、吴财、林湛
项目面积：140平方米
项目地点：浙江杭州
主要材料：大理石、地毯、墙纸等
摄 影 师：陈维忠

DESIGN CONCEPT / 设计理念

>> This case is prepared for the leaders of life. Whether the entrepreneurs or the outstanding people in the ordinary life, they are the leaders of this age. They are carefree in the prime of life and soaring leisurely in the sky. In such a changeable age, they hold fast to their dreams and explore their own world with open mind. I like the interpretation of "prime-time", which is not as bare as "mid-age", and "prime-time" means "golden period".

>> 本户型是为生活中的领航者准备的。无论是商界企业家还是平凡生活中的佼佼者，他们都是这个时代的领航者。他们逍遥壮年，他们乘风去来，任意翱翔在天际。在一个瞬息万变的时代里，他们将梦想坚持到底，瞬息百里，以超脱的豁达开辟出属于自己的一片天地。我喜欢英文中"中年"的代称，不是"mid-age"这么赤裸裸的直称，而是用"prime-time"，是"黄金时段"。

If you say, the prime-time is the watershed of one's life, then in this age, people will become mature, sensitive and decisive and will have bountiful gain. And this is also the time for thriving your dream, making a smooth "prime-time". In the prime time, they traversed lush forest, viewed the broad desert and experienced vast ocean and become more unhurried, just like the city of Hangzhou. After experiencing the integration of old and new civilizations, it still shows its unique luxury and elegance. We can see perseverance in its peace, aggressiveness in its gentility, calmness in its ordinariness, which is also the best appearance of what the residence wants to present.

若说 prime-time 是人生的分水岭，那么这个年纪在人生阅历中，的确是水到渠成的成熟，既敏感又果断，那种岁月换来的成熟，也使一切的收获都有着黄金的分量。而这也是人生阶段中最能事业有成的茁梦时刻，使得"prime-time"们有一种信手拈来的顺遂。逍遥壮年，他们穿越繁茂的森林、见过广袤的大漠、经历过浩瀚的大海，他们愈来愈从容，就如杭州这座城市一样，在经过新旧文明融合的洗礼之后，依然透露着自己独有的奢雅。在安安静静中透着坚韧、斯斯文文里带着一点点霸气、平平凡凡中透着一股淡定的气场，这也是这居所想要呈现的同样的绝佳面貌。

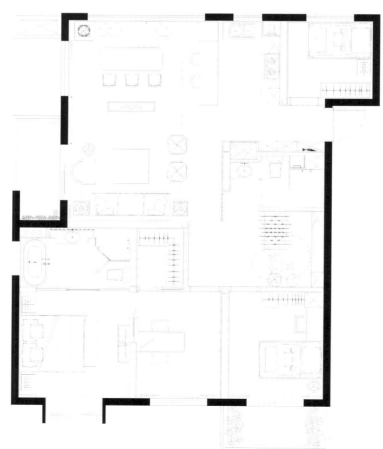

ART SPACE
艺术空间

设计公司：百搭园软装
设计总监：司蓉
主设计师：郭健、孟丹
项目面积：71平方米
项目地点：广东广州
主要材料：大理石、壁纸、木地板等

DESIGN CONCEPT / 设计理念

>> The entire space starts from gentleman's delicate and fashionable life pattern and combines the most advanced interior design to create the most top hotel-style apartment in Guangzhou with high-level panoramic French windows which offer broad views. On the basis of luxurious hard decoration, there is no excessive colorful or complicated soft decoration. In this way, each corner seems causal but without free attitude or expression, the gentleman's taste and fashion can be transmitted from this small place.

>> 整个空间从绅士的精致时尚的生活方式出发，结合当下最前卫的室内设计风潮，打造广州最顶级酒店公寓，挑高层全景式落地窗，视野宽广。在结合奢华硬装的基础上，软装陈设无需过多绚丽的装饰，也无需过多复杂的设计，让每个角落看似漫不经心却有着并非随性的态度与表现，让绅士的品位、时尚感都能从这片小小的方寸之地传递出去。

095

The high living room makes for a broad vision, the entire space is in simple but elegant color, Andorra's brown and gray adjust the tone of the space. At the same time, various materials are used to reflect the layering of the space, leather and cloth are used to synthesis the hard lines, making the coordination and transition more obvious. In the selection of furniture, Casa milano, a high-end furniture brand, is applied to show the dignity of the owner. The dotting of rose gold, dark reddish purple and red creates the elegant style of gentleman.

挑高的客厅让视野更加宽广，整个空间配色简单"低雅"，以安道尔棕与灰调节空间的冷暖、进退。同时利用多种材质体现空间的层次感，皮质和布艺来综合硬朗的线条感，协调过渡的作用更加的鲜明。在家具选取上有着强有力的表现，采用"Casa milano"家具品牌体现主人尊贵的身份象征。玫瑰金与酱紫红的点缀，更成为打造绅士优雅格调的首选。

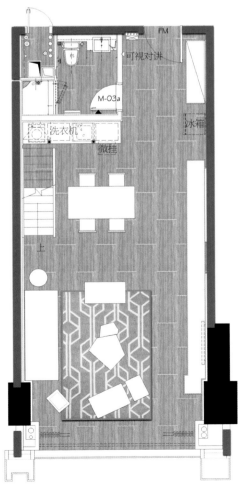

Following the staircase besides the bedroom and stepping into the second floor, you will see the private space for gentleman, which is simple, comfortable but with taste. The rose gold bed matches with the pure white soft cushion, which is clean and tidy and create a relaxed feeling for gentleman. In the selection of accessories, distinctive bedside cupboards are chosen to reflect the gentleman's attitude of refusal to mediocrity. As we can see from the small details in very corner, they refuse being mediocre, and pay attention to detail and represent "new gentleman" of city with handsome appearance, fashionable cloths, elegant taste and wise behavior.This case breaks the stereotype of the gentleman, and presents more interesting, fun and modern gentleman. The space illustrates the life bearing of "gentleman" and focuses on the coordination of living habit and the application of natural materials. The designers hope to use the simplest lines, shapes and most fashionable elements to show the understanding and love to life.

　　卧室沿着楼梯上二楼，便是绅士的私人空间，整个空间简约、舒适，又不输品位。玫瑰金床搭配纯白舒适的软垫，干净、整洁，让绅士回家有一种心灵释放的感觉。在物品的选择上不走寻常路，选用打破常规具有设计感的床头柜，体现新绅士拒绝平庸的生活态度。从卧室各个角落的小细节都可以看出，他们拒绝平庸，注重细节，是外表俊逸，衣着时尚，品味优雅，举止从容睿智的都市"新绅士"。此案打破了其以往塑造的既有绅士形象，让绅士们变得更有趣，更好玩，更摩登。空间陈述了一个"新绅士"都市人的生活风范，它注重人与生活习惯的协调以及对天然材质的使用，期望用最简约的线条廓形、最时尚潮流的元素，展示对生活的理解与热爱。

IMPRESSIVE SYMPHONY
印象交响乐

设计公司：上海迅美装饰设计有限公司
设 计 师：赵辉
项目面积：340 平方米
项目地点：吉林长春
主要材料：爵士白大理石、咖啡色大理石、肌理壁纸、金属收边、硬包、拼花地板、镜面等

DESIGN CONCEPT / 设计理念

>> This case merges both Chinese and Western styles, it is designed for a successful person working in a foreign company whose hobbies include tea ceremony, golf, meditation, music. The designer uses the collision of colors skillfully to create a warm and fashionable atmosphere. Besides, designer refers to the textures and colors of superfluity, making a delicate and modern space filled with luxury without losing the taste. While emphasizing the spiritual appearance of city, this case also pays attention to integrate traditional values. Artistic painting, Neo-Chinese style desks chairs, elaborately designed cupboard for books and pottery are chose in this case, which is not only practical but also shows the cultural deposition.

>> 本户型的定位是成功的外企人士，主人的爱好为茶道、高尔夫、冥想、音乐，风格定位为中西融合。设计师巧妙运用颜色碰撞，使气氛更加温馨时尚，另外借鉴了奢侈品的纹样和颜色设计风格，让整体空间呈现出一种奢华又不失品位的精致感与现代感。在着重体现都市精神面貌的同时还不忘融合传统价值理念，选用有艺术感的挂画、配置有线条感的新中式桌椅调配空间比例、精心打造展示书籍与陶艺品的壁柜等等，既实用又增加文化气息。

The entire space on the first floor is smooth and with full layout, designer made some changes in the staircase in the original living room, so that the living room can be more structured and magnificent. The living room includes reception area and fireplace area. The dining room continues the design of living room, with layered ceiling and elaborate marble mosaic pavement. The wall is decorated with the combination of marble and wood finishes, and it is designed in a sideboard form, which is beautiful and practical when matched with the European style wall lamp. The original two bedrooms are transformed into three bedrooms on the second floor to fulfill the demand of family members. The basement is designed in New-Chinese style, the original courtyard is now encompassed indoor, which adds its function. The trace of the host can be found in the calligraphy and painting area, tea area, pottery area and wine-tasting area. In addition, there is a yoga mediation room, which is a place for adjusting body, breath and heart.

一层整个空间动线流畅，布局饱满，更改了原建筑客厅处楼梯台阶，使客厅更为方正规整和大气，客厅分为会客区及壁炉区。餐厅沿用了客厅富有层次的天花及精致的大理石拼花地面，墙面用大理石和木饰面相结合，设计成餐边柜形式，配上欧式壁灯，既美观又实用。二层将原建筑的两个卧室改成了三个卧室，满足多人口之家之需。地下层的整个空间为新中式风格，将原建筑的庭院囊括到室内，增强了使用功能。书画区、品茶区、陶艺区、品酒区，都有主人的生活痕迹。另外还设置瑜珈冥想室，这是一个调身、调息、调心的空间。

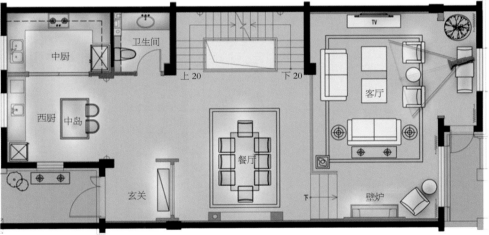

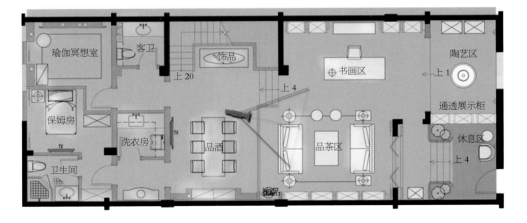

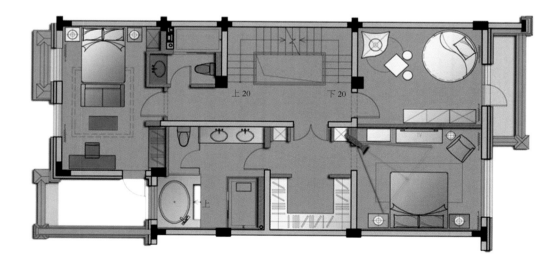

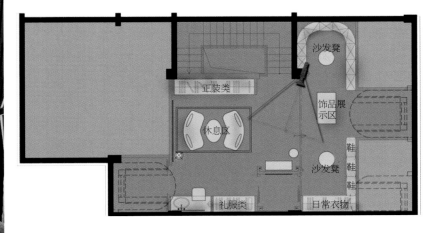

THE INTERWEAVE OF LUXURIOUS ELEGANCE AND MODERN SIMPLICITY

奢华典雅与现代简约交织

设计公司：西象建筑设计工程（上海）有限公司
设 计 师：何文哲
项目面积：485平方米
项目地点：浙江杭州
主要材料：大理石、吊灯、玻璃、木质地板等

DESIGN CONCEPT / 设计理念

>> There is a kind of beauty which is elegant and fresh. Light yellow, light beige and bright blue, these elegant colors paint a series of fashionable, gorgeous and vigorous scenes for this villa. Large scales of French windows make it possible for bringing enough light into the living room, so that the entire space is bathed in warm and sunny atmosphere. With the decoration of Kashmir wool, leather, cushion, cotton and wood, the living space forms an elegant unity. The light gray color matches well with the bright yellow sofa and modular retro yet modern tea tables, only the fashionable and cold colored carpet could overwhelm this low-key but magnificent momentum.

>> 有一种美，它优雅而清新。浅黄、淡米、亮蓝，淡雅的色调，为我们这个别墅空间描绘出一幅幅时髦典雅而充满生命力的场景。大面的落地窗让足够的光线穿透客厅，使整个空间沉浸在和煦明媚的氛围中。在克什米尔羊毛、皮革、衬垫、棉花、木头的诠释下，居家空间形成一种优雅自得的统一性。客厅里浅灰色搭配亮黄色的沙发，组合式的复古摩登茶几，唯有时尚的高冷系地毯才能压住这股低调而霸气的气场。

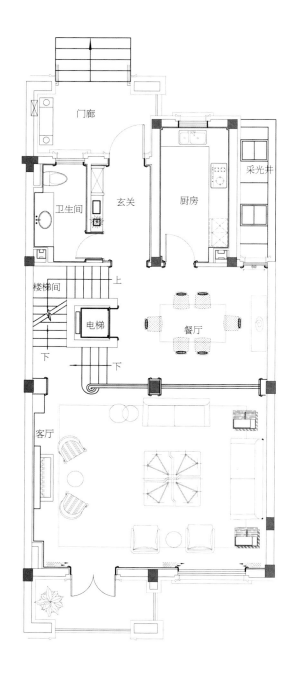
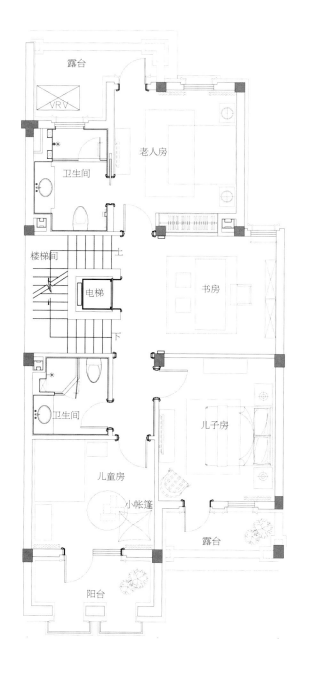

119

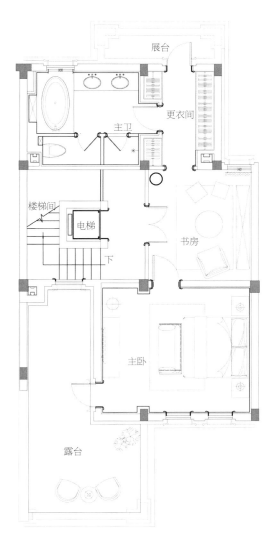

The family activity room creates a harmonious communication space for the elders and children by combining the functions, colors and hobbies. The designer uses the layout closes to life to create a living space for the family.

The master room is on the third floor, passing through the reception room, you will enter into the bedroom. The adequate light penetrates the gray linen fabric curtain and comes into the room, making it natural and harmonious. The calm color creates a warm and harmonious living environment together with the soft fabrics.

The interlayer in the basement provides an interactive space to rehearse and compose for the teenagers who love music.

On the underground floor, there is a high space, the designer applies bicycle riding and climbing themes into the space to create an extreme sports CLUB full of adventurous fun.

　　二层的亲子活动室从功能、色系、兴趣等几方面结合，为长辈与孩子营造了一个和谐的活动交流之处。用最贴合生活主题的布局方式来打造儿女们和长辈的居住空间。

　　主人房位于三层，通过会客室进入卧室，充足的光线穿过灰色系亚麻面料窗帘透进房间，显得自然而和谐。沉稳内敛的色系与充满温和气息的织物营造出一个温馨和谐的家居环境。

　　地下夹层为喜爱音乐的少年们提供了一个排练创作的互动空间。

　　地下一层有一个挑高的空间，设计师注入单车骑行和攀岩的主题，打造一个富有冒险趣味的极限运动CLUB。

现代奢华风格
MODERN LUXURIOUS STYLE

ART PARTY
艺术派对

项目名称：成都·伊泰天骄 370 户型样板房
陈设机构：深圳臻品设计
软装顾问：深圳华墨国际
项目面积：370 平方米
项目地点：四川成都
主要材料：大理石、木地板、油画等

DESIGN CONCEPT / 设计理念

>> It has the unique temperament of metropolitan; it has super size that never appeared in Chengdu; it only creates the purest lifestyle of city core. It seems to be the elegance in London, the fashion in Milan, the romance in Paris, the Luxury in Tokyo, the generosity in Manhattan and the passion in Spain. Its exalted, elegant and restrained flavor is like a mature and charming gentleman who is tasting the mocha coffee and thinking quietly. The soft decoration of this project adopts low-key and luxurious design concept and applies international and modern aesthetic vision. Reasonable and clear layout, fashionable and generous colors and exquisite and luxurious furnishings create a modern, luxurious and exquisite home.

>> 独一无二的大都会气质，拥有成都未曾出现过的超大尺度，缔造城市核心最纯粹的生活方式。仿佛伦敦的优雅、米兰的时尚、巴黎的浪漫、东京的奢华、曼哈顿的大气、西班牙的激情，尊贵、高雅、内敛的气息，犹如一位成熟而魅力的绅士细细品味着摩卡咖啡，静静地思索，悠远流长。本案软装陈设以低调奢华的设计理念，运用国际化与现代化美学的视觉审美，通过合理清晰的布局，时尚大气的用色，精致奢华的配饰来打造既具有时尚感又具有奢华感的精致之家。

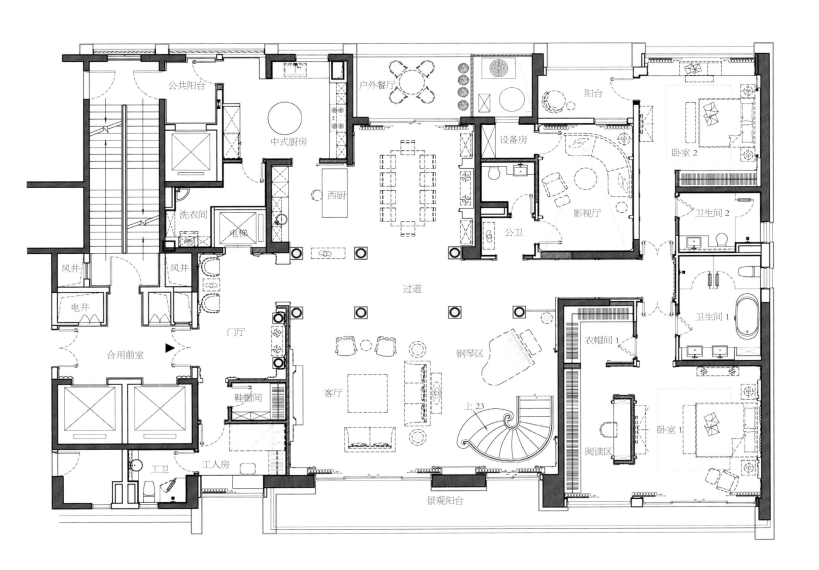

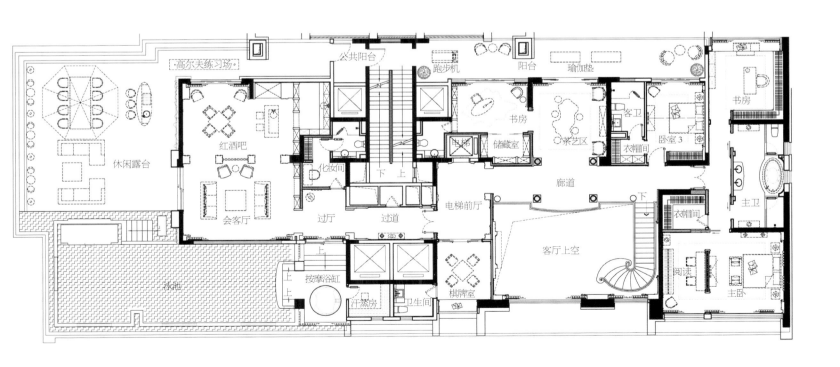

From the hallway to living room to dining room, the designer adopts open-vision layout. The color tone of the space is the combination of sapphire, black, white, gold and beige brown, which is charming and elegant and manifests noble flavor everywhere. Sapphire is used in many spaces. Blue is the soul of sea, is more profound than black, is more mysterious than purple, and is full of chance with infinite broad. The dining room stresses the dining order and etiquette. Solid wood dining chairs, beige brown leather dining chairs and exquisite flowers integrate in the space, which is natural and unrestrained, displaying an organic fusion of art and life.

从门厅到客厅再到餐厅，采用了视野开阔的开放式格局，颜色搭配把"宝石蓝色"、黑、白、金、米棕色作为空间的色彩基调，迷人而优雅，处处彰显贵族的气息。"宝石蓝色"穿插在多个空间，蓝色是海的灵魂，比黑色更深刻，比紫色更神秘，充满变化，无限宽广。餐厅强调用餐的秩序和礼仪，实木餐椅、米棕色皮质餐椅、精美花艺在空间中融合汇聚，自然洒脱，呈现艺术与生活的有机融合。

QUALITY LIVING
品质生活

项目名称：万科翡翠滨江 165 样板房
软装公司：DML Design 麟美国际陈设机构
设 计 师：董美麟
项目面积：165 平方米
项目地点：上海
主要材料：大理石、玻璃、木质地板等
摄 影 师：Hans Fonk, Aguang

DESIGN CONCEPT / 设计理念

>> Quality living is to make complicated and concrete life present artistic characteristic, at the same time, to reflect people's love for life and a fun life status. In the busy and complicated city life, this status makes "home" no longer a one-way street, no longer a house or a showy store, but a reflection of details in comfortable life. So when positioning the characters, the designers chose returnee couple who have their own beloved career and passion for life. The host is an upstart in finance and the hostess is a fashion buyer and florist. They are hospitable as the quality life lies in spiritual giving for them, so they like social activities and always share the details of life via home party.

>> 品质生活是让繁琐的具象的生活呈现出生活艺术的特征，又反映出生活中的人对生活的热爱和把握，以及对生活游刃有余充满乐趣的一种状态。游刃有余和充满乐趣的状态，在繁忙复杂的都市生活中"家"不再是个单行线，"家"不再只是房子或者炫耀的卖场，"家"是温馨的生活细节体现。所以我们定位人物时候选择了游学归国并同时拥有自己热爱的职业，热爱生活的夫妻。男主人金融新贵，女主人时尚买手花艺师，他们热情好客，因为对于他们来说生活品质的源头在精神的给予，所以他们喜爱社交，并经常通过 home party 分享生活的细节。

As for the colors, the designers chose coffee and silver color as the keynote, gray–blue is used throughout the space, the master bedroom is dotted with dark red. Furniture, handmade imported New Zealand wool carpet, oil painting and leather carving are designed with geometric figure and the distortion of floriculture symbol to enlarge the details. As for the handmade crystal chandelier and furniture, the designers chose white of volakas, delicate stainless steel hardware and embedded embellishment, which can reveal the high quality pursuit in rationality and spirituality.

　　设计的色彩，我们选择了咖色银色基调做主体，灰蓝色贯穿始终，主人卧室暗红色点缀。家具、手工进口新西兰羊毛地毯、油画、入口皮雕，我们选择几何图形和花艺符号的变型和放大着眼细节。手工水晶吊灯，家具选用爵士白大理石和非常精致的不锈钢五金收口及嵌入式点缀，体现男主人的理性和精神极致的高品位追求。

NIFTY MAGICLISM
俏皮魔幻主义

现代奢华风格 / MODERN LUXURIOUS STYLE

项目名称：电力家和城
设计公司：珠海空间印象建筑装饰设计有限公司
设 计 师：吕道伟
项目面积：661平方米
项目地点：广东珠海
主要材料：大理石、玻璃、地砖等

DESIGN CONCEPT / 设计理念

>> Entering this show flat is like stepping into the set of a fashion movie. The magnificent metropolis atmosphere, delicate neo-classical patterns, neat up-winding lamp ribbon are in nifty magiclism style, presenting the "personal" confidence. The designer wants to continue his personal fashion taste here. In every space, the clean and neat lines form into groups properly, scattering and gathering. The surface flows charm and elegance, while the connotation is rigorous space logic. The master combines the design language, shape and color and kinetonema together in the artistic conception, creating a fabulous stage play.

>> 走进家和城的这套样板别墅，仿佛步入时尚大片的片场，张扬的都会气氛，精致的新古典花纹，整齐的灯带蜿蜒而上，带着魔幻主义的俏皮风格，表现出"我行我素"的自信。设计师欲将其个性的时尚品位续存延续在此，各个空间以干净利落的流线结合群组，疏密有道，丝丝漫开并悄悄收拢；表面是流淌的风韵韶华，内涵则是严谨的空间逻辑思维；这位演绎大师将空间的设计语言、形色、动线做深浅意境之弥合，打造一场美妙绝伦的舞台剧。

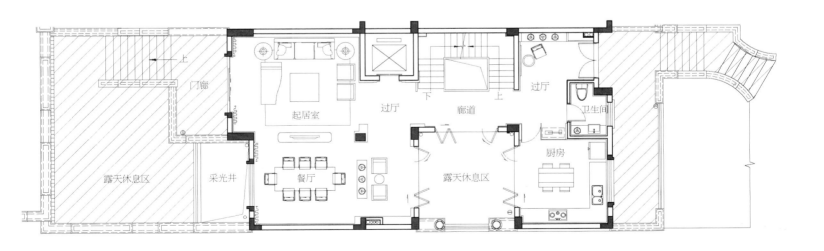

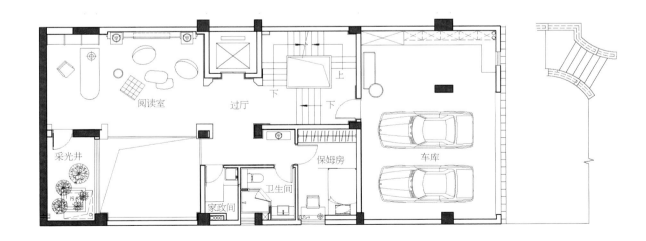

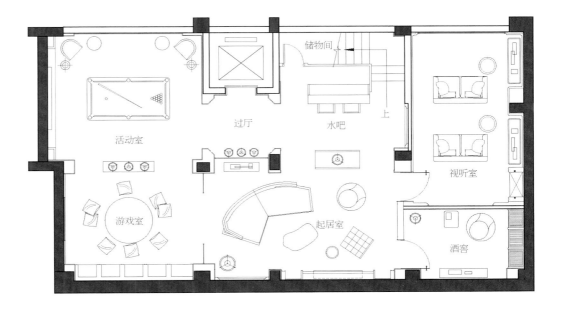

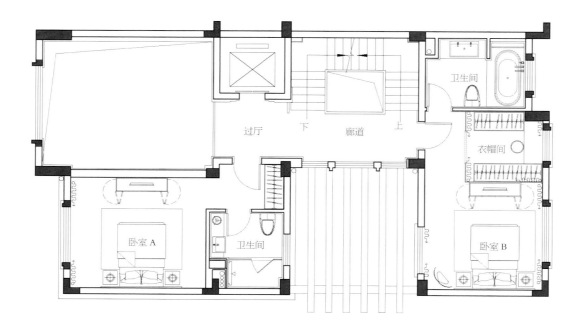

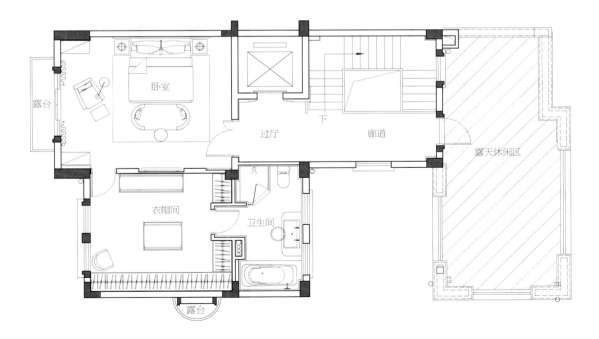

PAINTING THE PHANTOM OF CITY

绮绘都市魅影

项目名称：芜湖伟星金域蓝湾
设计公司：云庭设计（南京云庭家居艺术用品有限公司）
设 计 师：陈跃飞、赵静、袁贝贝
项目面积：200平方米
项目地点：安徽芜湖
主要材料：影木、微晶石砖、皮革、不锈钢、水晶等
摄 影 师：逆风笑

DESIGN CONCEPT / 设计理念

>> Fashion, advancement and quality are the descriptions of this case. It abandoned the luxury and advocates concise. The designers mainly use beige, black and camel and dot pure blue, yellow to form the color scheme of this case. The combination of ornament, painting and furniture shows the gorgeous color element of the space, at the same time, metal and marble are integrated to add the decoration and texture of the space. In this way, visual symbols are converted into the space to present a sensory space with modern simplicity and high-end quality.

>> 新锐、前卫、品质是本案的代名词，设计摒弃浮华，崇尚精炼。设计中，空间色调以米灰色为主色调，以简洁明快的设计手法对前卫、品位加以诠释，营造出独具风度的空间质感和视觉观感。陈设色调以米色、黑色、驼色为主，点缀纯正的蓝色和黄色，形成本案的气质色彩。通过饰品、挂画、家具细节等各处进行组合搭配，向整个空间传递华美的色彩元素；同时融入金属、大理石元素，增强空间的装饰性和品质感，形成一定的视觉符号转换到空间中，呈现出现代简洁，品质前卫的感观空间。

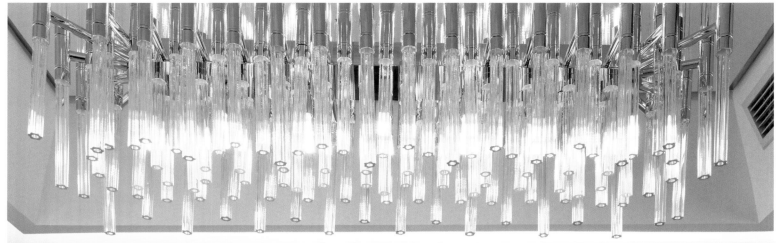

BRITISH & FRENCH STYLE 英法风格
MODERN LUXURIOUS STYLE 现代奢华风格
SIMPLE EUROPEAN STYLE 简欧风格
MODERN CHINESE STYLE 现代中式风格
MEDITERRANEAN STYLE 地中海风格

SIMPLE EUROPEAN STYLE
简欧风格

NEW INTERPRATION OF FASHION

时尚新解读

设计公司：Studio.Y 余颢凌设计工作室
室内设计师：张译丹
软装设计师：杨超
项目面积：245 平方米
项目地点：四川成都
主要材料：大理石、地砖、布艺等

DESIGN CONCEPT / 设计理念

>> The owners of this case are keen on modern and bright elements. They are as good as professional designers in the acuity of color and sensory of cloth. The pieces in the living room almost use arc-shaped or round-shaped or curved furniture, linear elements can be seen easily in this area. The parallel, flowing, hale, and assorted lines are vertically and horizontally intertwined, forming a distinctive fashionable space. Some adjustments are made on the wall of the living room, and special modelling disposal is added on the wall, imported Belgian HW and Italian HARLEQUIM wallpaper inject some artistic atmosphere.

>> 本案的业主酷爱一切摩登亮丽的风尚元素。对于色彩的敏锐度以及布艺面料的质感要求，他们不比专业设计师逊色。客餐厅的单品几乎全部采用弧形、圆形或曲线的造型，线条元素在这个区域一眼可见，平行的、飘逸的、硬朗的以及组合的，纵横交织，形成别具一格的空间摩登感。客餐厅的墙壁在原有的基础上作了调整和改造，在墙面加入特别的造型处理，采用比利时 HW 和英国 HARLEQUIM 品牌全进口独具质感的肌理墙纸，极富艺术气息。

Black, white and gold are the keynotes of this case, yellow and blue, the two beloved colors of the owners are added. Artistic special decorations, such as the Hermès silk scarf is framed into an amazing painting skillfully in the hallway, which is a salute to fashion and classic. In addition, famous artist Qu Guangci's precious art series *Angel–Qian* enlightened the space, which is the unique point of the designer.

整个案例以黑白金为主色调，加入业主酷爱的黄蓝撞色。极富艺术感的特别装饰，譬如进门玄关处，巧妙地用一条 Hermès 丝巾装裱成一幅绝佳挂画，无疑是对时尚经典的致敬。加之出自著名艺术家瞿广慈之手的稀奇艺术系列《最天使——乾》，让这个空间颇为亮眼，更是设计师的独具匠心之处。

THE JUMPING COLORS
跳跃的色彩

设计公司：上海益善堂装饰设计有限公司
主案设计：王利贤
参与设计：李丝莲、马玲玲
项目面积：220平方米
项目地点：江苏淮安
主要材料：玫瑰金不锈钢、白色高光漆木饰面、硬包、西洋米黄石材、黑白根大理石、橄榄灰大理石等

DESIGN CONCEPT / 设计理念

>> After taking the compact space and the young owner into consideration, the designers choose fashionable modern simple style in this case. They hope to satisfy the need for space environment in a simple way. Black, white and gray colors are mainly used in this case, proportional mango yellow and cyanid blue are boldly used, purple and red are dotted, so the space is in bright and fashionable colors. It inherits the simplicity and comfortable feeling of classic space, and it also creates different dynamic living experience.

>> 考虑到空间大小比较紧凑，受众群体相对年轻，设计师将其定位为时尚现代简约风格。以简洁的表现形式来满足人们对空间环境的需求。空间多以黑白灰表现为主，设计师大胆加入多比例的芒果黄色和车菊蓝，紫色和红色作为点缀，色彩明快而摩登。既延续经典的空间简约舒适，亦打造出不一样新鲜活力的生活体验。

Stepping into the entrance, you will see the reading area. In this contemporary society with fast rhythm, many people try to search for their peaceful experience by reading. In this area, the elk-themed wall decoration and bookshelf create a tranquil feeling of living harmoniously with animals in the nature and enable people read books peacefully in the evening. The reading area which is close to the entrance makes owners relax themselves the moment they return home.

进入大门后,迎面而来的是阅读区。在现代社会的快节奏下,很多人会通过阅读来寻找一份属于自己的安宁体验。在这个区域,以麋鹿为主题的墙饰和书架,营造出一份大自然中与动物和谐共处的宁静感,帮助读者进入安静的阅读环境。一个带有橡树果部件的黄铜落地灯,能让读者在夜晚安心品阅书籍。这个紧接入口的阅读区,能让主人在回家的那一刻,就放松自己的心情。

TONE
格调

设计公司：重庆星翰装饰设计工程有限公司
项目面积：351平方米
项目地点：四川成都
主要材料：不锈钢、大理石、木地板等
摄 影 师：重庆麒文建筑与空间摄影公司/张骑麟

DESIGN CONCEPT 设计理念

>> The designer of this project adopts white as the main tone and purple as the background, embellished with blue furnishings. The whole space presents a modern and romantic style. The living room is equipped with modern European sofas in different forms. The back of the sofas continues traditional symmetrical design. The comparison between leather and wired glass is well-designed. The fireplace combines light color stone with unique chic stainless steel lines, which is fashionable and exquisite. There are large areas of imported dark coffee marbles, embellished with gold flowers, being the punchline of the living room.

>> 本案设计师采用白色为主调，紫色为背景，配以蓝色饰品点缀，整体空间体现时尚、浪漫的格调。客厅选用现代欧式不同形态的沙发组合，沙发背幅延续传统对称设计，扣皮与夹丝玻璃两种质感对比，设计感十足。壁炉采用浅色石材搭配别致的不锈钢线条，在时尚中更精致，主幅采用进口深啡大块面呈现，以金色花点缀，成为整个客厅的点睛之笔。

The dining room is simple with detailed designs. The cabinet meets the needs to collect and display. Blue furnishings echo with the living room. The modeling of lines and surfaces is the continuation of the fashion style. The back wall in daughter's room chooses bold color. Other furnishings echo with purple as if the flowers burst into bloom with romantic tastes. The mirror cabinet in the bathroom has a sense of extension while the corrugated cabinet is simple and fashionable.

餐厅简约而不失细节，整面装饰柜满足收纳与展示，蓝色饰品与客厅遥相呼应，线面造型层次分明，正是时尚风格的延续。女儿房床背幅大胆跳色，其余饰品与紫色相呼应，如花开绽放般散发着浪漫的气息。卫生间镜柜有延伸空间感，波纹板柜体简洁时尚。

The son's room uses One Piece as the theme, which echoes with the blue tone of the entire design. The activity room which is injected with spirit of adventure is used as the children's public area, where purple and blue are harmonious and unified. The master bedroom continues the romantic and fashionable style in living room. The variations with blue ornaments and gold furnishings present the beauty of fashionable and romantic tastes. The study is decorated with special-shaped sofas, which is relaxing and comfortable. The symmetrical bookcase and disordered furnishings are chaos orderly. The design of the flexible and rigorous cloakroom meets the needs to collect and display. The bathroom boldly adopts black stone, complemented with gold stainless steel, which manifests the fashion taste of the owner. The cabinet is in arc-shape and wave-shape, which is avant-garde and elegant. The bar with red wine and cigar is decorated with dark color and makes a sharp contrast with the ceiling and floor. The whole space is luxurious. Partial blue ornaments carry romance through to the end.

儿子房以海贼王为主题，呼应整个设计蓝色点缀的色调，更将整个空间注入勇敢冒险的精神元素。活动室作为儿女的公共空间，紫色与蓝色和谐统一。主卧室沿袭客厅浪漫、时尚的风格，将蓝色点缀分出各种变化，局部饰品融入金色，是体现时尚浪漫品味的唯美注脚。书房以异形沙发布局，轻松舒适，书柜对称设计，饰品错落摆放，乱中有序，灵动而不失严谨。衣帽间设计巧妙满足收纳与展示。卫生间大胆采用黑色石材，配以金色不锈钢，彰显主人时尚的品位，柜体以弧形波浪形呈现，前卫优雅。红酒雪茄吧采用深色墙面，与天花板地面形成鲜明对比，整个空间弥漫着奢华的气质。局部蓝色点缀，将浪漫进行到底。

EUROPEAN STYLE

欧式风情

设计公司：浙江迈斯建筑装饰设计有限公司
软装设计：上海范尼斯特软装工程设计有限公司
设 计 师：陈丰华
项目面积：650 平方米
项目地点：浙江温州
主要材料：木饰面、大理石、多层实木地板、乳胶漆等

DESIGN CONCEPT / 设计理念

>> This private villa has five floors including the basemet. The interior is modern style with some European elements. The designs pursue clean and bright sense of space, which is concise and elegant. In order to create a light luxurious and elegant living tonality, the designer chooses the inclusive white as the basic color of the space and properly uses elegant materials and light tones to create a sense of texture. The white stairs with good modeling and the warm color increase the temperature of the space. The exquisite droplight enriches the expressions in the living room.

>> 本案私人别墅共五层（含地下室），室内风格以现代为主，融入了一些欧式元素，设计追求清澈明亮的空间感，简约、素雅。为打造一种轻奢、优雅的居家调性，设计师利用十分具有包容性的白色为空间的基底，以素雅的材质、轻盈的色调合宜规划，打造出质感空间。白色的造型感极强的楼梯，温润的颜色增强了空间的温度，细腻的吊顶也让客厅的表情变得更加丰富。

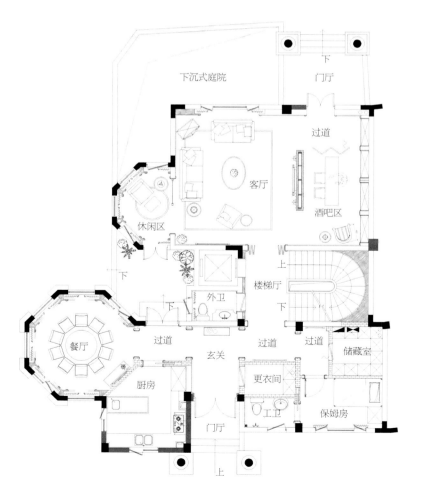

After a busy day, people may long to the natural, sheer, comfortable and romantic lesuire time. The European style of the space decreases the original European dignity and elegance and increases the natural and tranquil freshness and comfort. The details of the space focus on psychical return to nature. The wall is painted with fresh white, in addition with lively warm color furniture and sedate iron light, giving people a relaxing and comfortable feeling.

在结束一天繁忙的工作日程后,会更加憧憬自然纯粹舒适浪漫的闲暇时光。空间的欧式设计风格,少了一点欧式原有的高贵雍容,增添了自然恬淡的清新舒适。在空间的细节上讲求心灵的自然回归感,清新的白色漆上墙面,配上明快的暖色家具与沉稳的铁艺灯,给人放松惬意的舒适感。

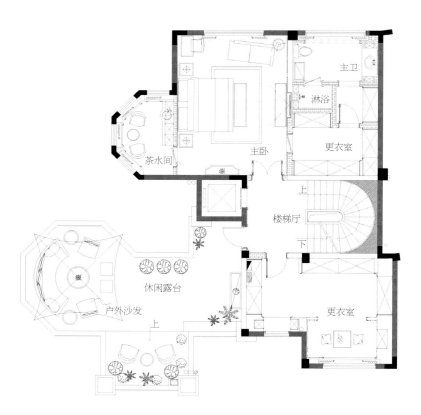
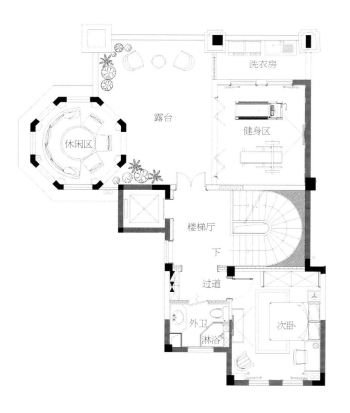

A RESIDENCE WITH SUNNY VIEW
向阳而居

设计公司：上海益善堂装饰设计有限公司
设 计 师：汤玉柱、马玲玲
项目面积：410平方米
项目地点：上海
主要材料：黑色镜面不锈钢、黑檀木木饰面、黑檀木木地板等
摄 影 师：温蔚汉

DESIGN CONCEPT / 设计理念

>> This case has top class river scenery with amazing view that can not be replicated. Based on light, happy and neutral conceptions, it applies modern style as the keynote. On space planning, it is not restrained to the limitation of traditional symmetry and prefers to free, open and innovative visual sense. Stainless steel, marble and other materials with era sensation are mainly used in this case, the main areas are dotted with bright colored ornamentations. The open and penetrable space as well as the personalized linear design makes for a space with smooth layout and natural integration. In this open environment, the fashionable furniture shows distinctive temperament.

>> 坐拥一线江景，绝美视野无以复制。本案围绕轻盈、欢快、中性的构思，以现代主义风格为主调，在空间平面设计中不受传统对称限制，追求自由开放、独具新意的视觉感官。重点采用了不锈钢、大理石等材料凸显时代感，主要区域搭配色彩浓烈的装饰加以点缀。空间开敞、内外通透，个性化的线性设计，使得空间布局流畅，各区域自然融合。时尚的家具在开放的环境下，彰显出与众不同的气质。

The designers pay much attention to showing the individuality and cultural connotation, advocating personality and opposing plain and stereotyped design. They use moveable elements, rich colors to improve the cultural connotation of the space and create a distinctive space with taste. The atmosphere in this space is rich and diverse, including golden and black colors which reveal solemn bearing, the elegance of blue color and Oriental sentiment.

设计师非常注重体现个性和文化内涵，在设计中强调人的个性，反对苍白平庸和千篇一律，体现个性化需求。通过可移动的元素，丰富的色彩，增加空间内的文化内涵，为本项目打造了有品位、有特色的空间。空间的氛围丰富且多元，金与黑呈现的庄重气度，蓝色的典雅，还有极富东方气息的情调。

SUNNY RESIDENCE
阳光住宅

设计公司：上海元柏建筑设计事务所
设 计 师：史迪威
项目地点：广东深圳
项目面积：192.5 平方米
主要材料：地毯、大理石、木作等

DESIGN CONCEPT / 设计理念

>> The project is in modern and fashionable style with arc triangle modeling. The gallery is interconnected with good lighting and ventilation. The building highlights small base which makes the round building more dynamic and flexible. The house type is a big house in the building. The entire style is comfortable and generous. The transparent house type is the most shocking one in the building. In half-arc shape, the house faces south and sun with good lighting. Being here, one can enjoy the beautiful natural scenery.

>> 整体现代时尚风格，以弧形三角形为造型，内廊贯通，360度采光通风，彼此高低结合，楼体以底部见小为亮点，也让圆楼设计更加动感而不死板。本户型是该楼盘的大户型，整体风格让人感觉舒适大气，通透的户型设计感觉是整个楼盘最震撼的空间。该户型整体以半弧形设计打造，正南朝阳，采光极好，身在其中，通透明亮，可以欣赏秋水共长天一色的自然美景。

There is a bedroom in the lower lever, while there is a luxurious master bedroom in the upper with a study and a bathroom. The luxurious space makes the residents enjoy the inclusiveness of the home.

复式下层设有一间卧室,而上层则是奢华主人房,配置书房与卫浴,奢侈的空间尺度让居者尽享家的包容。

THE COLLISION OF ART
艺术的碰撞

设计公司：上海亚邑室内设计有限公司
设 计 师：孙建亚
项目地点：上海
主要材料：橡木、白色大理石、古铜等
摄 影 师：孙建亚

DESIGN CONCEPT / 设计理念

>> In the present market, for the high-end house buyers who have viewed the wide world, the luxury decoration is not the suitable choice for returning to life. But while maintaining the high-end material life, they also need inner life taste, and begin to pursue different spiritual need and styles. For this kind of pursuit, the designer in this case chose different design method. He used simplified vertical plane ornament, polished dark wood veneer to match with the luxury bronze metal lines, which replaced the traditional luxury and complicated shape. The color temperature, material, method are unified, creating highly harmony between the component elements and furniture.

>> 现今市场，对于看尽大千世界的高端主力购房者，奢华装饰并不是当下回归生活的合适选择，但保留高端物质生活的同时却需要有内在的生活品位，开始转变张显各自不同精神需求与风格特色。对于这方面的需求，我们对空间设计的手法与以往有些不同，通过简化的立面装饰，高光深色雀眼木皮，搭配具有奢华感的古铜金属线条，替代了传统奢华繁琐的复杂造型。空间色温、材料、手法统一贯穿，构成元素与家具取得高度和谐。

BRITISH & FRENCH STYLE 英法风格
MODERN LUXURIOUS STYLE 现代奢华风格
SIMPLE EUROPEAN STYLE 简欧风格
MODERN CHINESE STYLE 现代中式风格
MEDITERRANEAN STYLE 地中海风格

MODERN CHINESE STYLE
现代中式风格

THE ANNUAL RING
年轮

设计公司：矩阵纵横
设 计 师：王冠、刘建辉、王兆宝
项目面积：225平方米
项目地点：广东深圳
主要材料：黑镜钢、爵士白大理石、灰木纹大理石、墙纸、实木地板、艺术玻璃等

DESIGN CONCEPT / 设计理念

>> The whole style of this project is sedate Chinese style, which interprets the generous spirit and noble and classical taste of life. The combination of modern and traditional elements, classical flavors created by modern aesthetics and concise and generous techniques present the owners a classic and gorgeous residence which conveys their elegant and comfortable attitude towards life. Proper treatments of structure and modeling aim at creating an elegant space full of cultural connotations.

>> 本项目整体风格是用中式而稳重的手法演绎大气精神，同时又不失高尚古典的品位生活。将现代和传统元素结合在一起，以现代人的审美打造古典韵味的事物，运用简约大气的手法，为业主营造一处经典、气派的大居所，体现出主人优雅舒适的生活态度。在空间与造型的处理上，尺度的把握恰到好处，宗旨在营造一个蕴含文化气息的雅致空间。

The whole layout meets the needs of life. The materials are mainly marbles and solid woods. Noble and sedate walnut and simple and generous wallpaper are decorated in the walls and partitions. The alternate applications of all kinds of Chinese elements present a modern and concise sense, creating a warm and elegant living environment for the owners.

在总体布局上尽量满足生活的需求，装修材料以大理石、实木为主，以胡桃木的尊贵稳重，壁纸的朴素大方来装饰墙面及隔断，各种中式元素的穿插应用更体现现代简约之感，为居住者创造一个温馨，优雅的家庭环境。

MODERN NEW HOME
时代新居

设计公司：上海元柏建筑设计事务所
设 计 师：史迪威
项目面积：175 平方米
项目地点：上海
主要材料：地砖、地毯、大理石等

DESIGN CONCEPT / 设计理念

>> The Neo-Chinese style was born in the new period of Chinese traditional cultural revival. With the strengthening of national power and the gradual recovery of national consciousness, people start to sort out the cue from the troublous "imitation" and "copy" to dig up a new era significance of Chinese traditional designs. At the beginning of exploring the native consciousness of Chinese design, a gradually mature design team and the consumer market breed the graceful Neo-Chinese style.

>> 新中式风格诞生于中国传统文化复兴的新时期，伴随着国力增强，民族意识逐渐复苏，人们开始从纷乱的"摹仿"和"拷贝"中整理出头绪，挖掘出中国传统设计的新时代意义。在探寻中国设计界的本土意识之初，逐渐成熟的新一代设计队伍和消费市场孕育出含蓄秀美的新中式风格。

This project captures the era characteristics of Chinese culture which sweeps around the world, ingeniously incorporates Chinese elements with modern materials to make the classical furniture, window lattices and furnishings complement with each other and reproduces the delicate items.The project inherits the essences of living concepts in Ming and Qing Dynasties, refines and enriches the classic elements, changes hierarchical and ranking feudal thoughts in the original layout, injects new flavor into the traditional living culture and achieves the perfect integration between tradition and modernity.

　　本案抓住了中国文化风靡全球的时代特点，将中式元素与现代材质巧妙兼柔，使古典式家具、窗棂、布艺床品相互辉映，再现了移步变景的精妙小品。本设计继承明清时期家居理念的精华，将其中的经典元素提炼并加以丰富，同时改变原有空间布局中等级、尊卑等封建思想，给传统家居文化注入了新的气息，实现了传统与现代的完美相融。

LOOKING FOR ORIENTAL FLAVOR : A SECRET FRAGRANCE

暗香·寻觅东方韵

设计公司：上海 ARCHI 意·嘉丰设计机构
设计团队：陈丹凌、彭兴、丛晓岑、罗剑、刘仲燕
项目面积：398 平方米
项目地点：吉林长春
主要材料：黑白根理石、夜来香理石、酸枝木、烤漆板、金箔漆、中式元素壁画等
摄　　影：空间摄影工作室

DESIGN CONCEPT／设计理念

>> In the whole space, the applications of marbles, rosewood, painted veneers and dragon plates set a fashionable keynote for the space. Wall paintings with retro Chinese elements, Zen-like porcelains and potteries and Neo-Chinese furniture with simple modeling add quaint flavor to the space. In color planning, warm gray is the basic color and peacock blue is the subject color, making the quiet and gentle space more flexible and vivid. With "a secret fragrance" as the theme, the fragrances of ink, flower, tea and incense interpret an extraordinary modern Chinese space, letting people regain the ancients' elegance and freedom and look for the grace and delicacy of Oriental flavor.

>> 在整个空间中，大理石、酸枝木、烤漆饰面、帝龙板等现代材料的运用，为空间奠定了富有时代感的基底；复古的中国元素壁画、禅意的瓷器和陶艺、造型简洁的新中式家具又为空间增添了古韵精髓，色彩的规划上则以暖灰色为基础色，孔雀蓝为主题色，使原本安静温婉的空间更加灵活生动。以"暗香"为陈设主题，糅和墨香、花香、茶香、熏香四种味道元素，演绎了一个别具内涵的现代中式空间，让人重拾一份古人的飘逸与洒脱，寻觅东方风情的那份雍容与雅致。

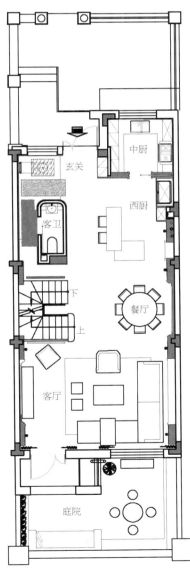
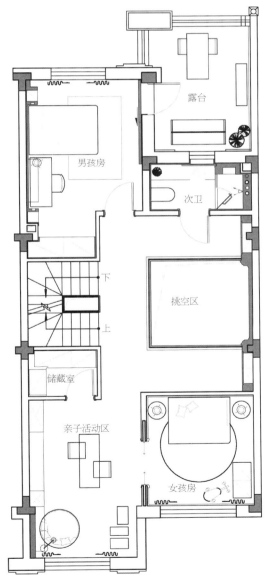

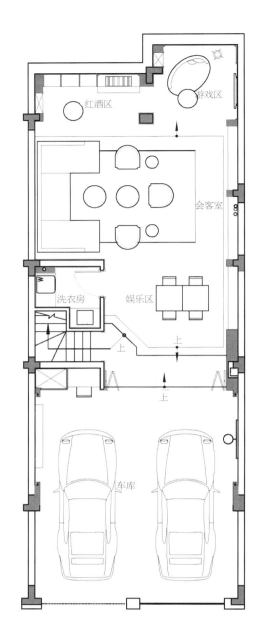

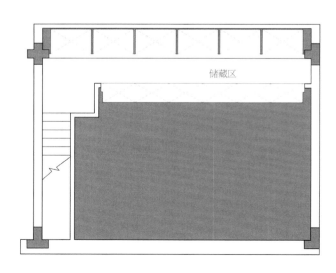

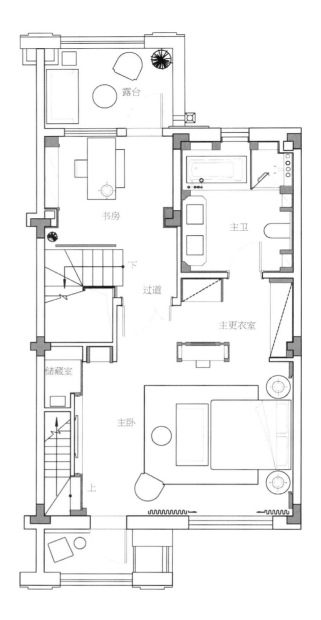

This show flat regards creating "a contemporary Chinese style space" as the main line, skillfully combines Chinese symbols with modern aesthetics, gets rid of concrete modeling, refines the traditional elements in Chinese culture, interprets by modern techniques and injects humanistic feelings into real life.

此样板间以打造"当代中式空间"为设计主线，将中式符号与现代审美巧妙融合，去掉具象的造型堆砌，将中式文化的传统元素精简提炼，再加以当代手法演绎，将人文情怀注入真实生活。

ATTACHMENTS OF BEAUTIFUL SCENERIES, POSSESSIONS OF TIMES

眷红偎翠，私藏岁月

设 计 师：连自成
软装设计：李曾娜、方丽云
项目面积：360平方米
项目地点：上海
主要材料：米白洞石、意大利木纹石、彩云飞大理石、胡桃木饰面、夹丝玻璃等
摄 影 师：隋思聪

DESIGN CONCEPT / 设计理念

>> For the designs of this space, designers want to absorb more elements from Oriental history and culture. There are many absorbable elements in the Chinese long history and complicated culture. At the same time, their purpose is to achieve a design which can make the living environment spread for a long time as the ancient trees. It seems to be a European residence where you can see traces of time, stories, cultures and emotions that the show flat is lack of. This is the psychological monologue of the designer Jacky Lien. Cultures can be strengthened and revived by architectures. He focuses on the ingenious spirits and elaborate degrees in ancient China and the contrast made by the modern machine manufacturing in the rapid consumption age.

>> 这一套空间的设计，更多想从东方的历史人文中抓取一些元素，中国历史的悠久，文化繁复错综，可汲取的元素很多，同时，我们的目的就是要一个设计，让空间居住的环境，和百年树海一样流传很久，就像欧洲的住宅一样，你可以看见岁月的痕迹，有故事有文化在里面，富有感情，而样板房缺少的其实就是感情，这是设计师连自成的心理独白。文化可由建筑凝固，同样可由建筑复苏。连自成注重那种中国古代的匠心精神，精细的程度，在快速消费时代下，与现代的机器制造形成反差。

The designers choose German porcelain brand as the leading role of space decorations, which not only pays respects to Chinese culture, but also interprets a different appearance of China by the Western design methods. In the designs of the living room, the traditional elements, such as sofas, tea tables, pillows with dragon patterns, silvered paintings with Kio and jumping color of Chinese red, stand out among the black and white furniture and lamps. The fusion of traditional and modern elements makes the space elegant and extraordinary.

设计师选择德国瓷器品牌作为空间装饰的主角,不仅致敬了中国文化,也借由国外设计师西化的手法演绎出别样的中国面貌。在客厅的设计中,沙发茶几以及龙纹图案的靠枕、锦鲤的银箔版画、跳跃的中国红,这些传统元素在黑白色的家具灯具的简洁中跳脱出来,这种传统与现代的混搭,让整个家中都找不到不够格调的平庸死角。

In addition, color is the biggest basis to review the style of the space as the classic color scheme in the French flag. Red and blue in low saturation combine with bright ivory white, light yellow and gold, which forms the color perceptions of the classic French family. And the perceptions are interpreted perfectly in the design of the bedroom by the designer Jacky Lien. Home is to be inherited. From the national culture to the comfort, here inherits many stories, such as growth, humanity and interest. People need to get along slowly with the house to produce emotions. When the house gets old, the feelings of being with old friends will come out.

除此之外，色彩是审视空间风格的最大依据，如同法国国旗上的经典配色，饱和度偏低的红与蓝，结合明度较高的象牙白、鹅黄或金色，组成了经典法式家庭的色彩感受。而这些感受被连自成完好的演绎于卧房的设计中。家是被传承的，从大的民族文化，到小的温馨，这里承载着很多故事：成长、人文、喜好。人和房子需要相处，慢慢的，才能滋生感情，等到房子有些旧了，那种老朋友的感觉就出来了。

现代中式风格 MODERN CHINESE STYLE

ORIENTAL CHARM
东方韵味

设计公司：北京意地筑作装饰设计有限公司
设 计 师：连志明、张伟
项目面积：154平方米
项目地点：北京
主要材料：水墨地毯、顽石、花草、水晶吊灯等

DESIGN CONCEPT / 设计理念

>> Leisurely and carefree mood is a fashion for literati in ancient times, but in a society with fast pace, noise and vanity, people still have deep yearning for "Building a house near where others live, but with no sound of wheels or voices". Home, a private living space is a place to release such a feeling. This case brings such a soft home. Oriental charm always glitters mysterious light, and the whole world never cease its exploration. The twenty-first century is an age that advocates humanization, it accepts humanization and the expression of feelings. Now, the classical beauty meets modern beauty, which unscrambles the world's resonance to Oriental aesthetics. The Oriental elements have swept the interior design world quietly. The designers combine Oriental charm and Western charm perfectly, showing a high-quality life.

> 闲情逸致是古时文人雅士的时尚，而当今生活在这快节奏、喧嚣浮华世间的人们对"结庐在人境，而无车马喧"仍留有那份深沉的向往。家，这个生活的私密空间即是能释放这种情怀的梦柔之乡，本案则将这梦柔之乡带入人境。东方韵味始终闪烁着神秘的光芒，全世界也从未停止对其探寻的脚步。21世纪是一个主张人性化的时代，认同人的个性化和个人情感表达。而今，东方古典美感与现代美感重新相遇，诠释出全球对东方审美爱好的共鸣。东方元素的设计悄悄席卷了室内设计界。设计师将东方的情调、西方的韵致完美地结合在一起，展现了一种当下高品质生活的诠释方式。

In this age of cultural integration, for Chinese designers, Oriental elements mean cultural heritage and the exploration of brilliant history, and it is also the foundation and the origin in design. In this case, we may see some Chinese style details, such as ink painting carpets, rocks, flowers, grass and Chinese painting, with this typical traditional Chinese marks, the designers interpret Oriental culture in a new way and they apply them these marks in the collocation of interior design, by combining humane languages, they show the modern and traditional charms vividly. This case shows a lasting charm without noise or complicated decorations. A tranquil Oriental style with luxury hidden in its details.

在东西文化大融合的时代，东方元素对于中国设计师而言，不仅是肩负起对中华民族文化传承的责任，探寻中国沿袭的千年文化；更是本土设计师在设计过程中寻宗归本的根脉。本案中可以看到很多中式的细节，譬如水墨地毯、顽石、花草以及国画，通过这些带有浓郁东方传统特色的符号，设计师运用一种全新的语言诠释东方文化，并将其运用到室内设计空间的搭配中，再结合人文气质的符号语言，把兼备现代与传统的东方韵味发挥得淋漓尽致。本案在透露出一种古色古香，韵味悠长之余，没有喧嚣与繁缛。一派宁静致远的东方式生活，细节之下浅埋着奢华。

A CHARMING NEW HOME IN SUZHOU
苏韵新居

设计公司：元禾大千
设 计 师：吴梁祝、耿波
项目地点：江苏苏州
主要材料：地毯、壁画、水晶吊灯等
摄 影 师：关晴飞

DESIGN CONCEPT / 设计理念

>> It is said there is a paradise in heaven, while there are Suzhou and Hangzhou on earth. The beautiful and elegant Suzhou Garden is called "the best in Jiangnan District". The bridge and water are surrounded by each other in Suzhou City. Suzhou Pingtan shows its ancient tones. The famous cultural city is significant for Chinese people. One novel part of this project is the Chinese neo-classical style with dark blue and bright yellow. Warmth and calmness permeate into each other, colliding a new visual effect and giving people a comfortable and fresh living space.

>> 正所谓"上有天堂，下有苏杭"，苏州园林秀丽典雅有着"甲江南"之美名，姑苏城小桥流水柔媚环绕，苏州评弹咿呀婉转吟唱着古老曲调，这个文化名城一直都是中国人无法忽视的。本设计的一个新颖之处在于，在中式新古典中加入了深蓝和亮黄的跳跃色彩，热烈与冷静相互渗透，碰撞出了一个具有全新视觉效果并给人舒适愉悦且带新鲜感的家居空间。

The decorations of the space focus on details, such as modern lacquer painting with birds and flowers, translucent crystal droplight and Chinese carpet. The designers put the Chinese elements in them to connect the whole space and create a harmony of living atmosphere and traditional art. The colors and patterns of the carpet and wall painting are chic. The white hollowed-out wall inspired by the classical garden pane type and the ceiling bring natural lights. Natural lights match with appropriate colors, creating a flavor of Neo-Chinese fusion style, which is interesting. The fusion comes from designers' thoughts to independent spaces.

在空间装饰上着重细节，现代花鸟漆画、晶莹剔透的水晶吊灯、浓厚的中式地毯，设计师将中式元素贯穿其中，衔接整个空间，营造出一种家居环境与传统艺术的和谐。地毯和壁画颜色图案别致，以古典园林窗格样式为灵感的白色镂空墙体和天花板带来自然光线。自然光线配上色彩的布局，产生一种新中式混搭的味道，妙趣横生，这种混搭源自于设计师对空间创新独立跨界思考的结果。

New Chinese pictures are traditional and novel, colliding an impressive Neo-Chinese space. The calm and elegant blue and white porcelain elements, the heavy old-fashioned wooden armchairs, the ubiquitous landscape elements, the bold colors and the modern crystal droplights are the highlights of this project.

新派中式挂画，传统与新颖相映成趣，碰撞出一个让人印象深刻的新中式空间。冷静与儒雅气质的青花瓷元素、厚重的太师椅、无处不在的山水元素以及大胆的色彩、现代感的水晶灯，都是本设计空间的亮点所在。

THE MEMORY OF PATIO
天井的记忆

设计公司：上海 CFC 环境设计机构
设 计 师：成志、胡俊峰
项目面积：2000 平方米
项目地点：上海
主要材料：土建遗留余土、回收旧木等
摄 影 师：张静

DESIGN CONCEPT / 设计理念

>> This is used to be a dumpsite in the eastern suburbs of the city, it took us three years to plant trees and recover the small scar of the earth. We build ten apartments, so that people could stay and talk here. It is fortunate that people will talk about topics about human and nature in such an environment, which is what we expected most. What is the most suitable relation of human, nature and architecture? In my memory, when I was a child, the yard surrounded by the houses around in my grandfather's house is the unforgettable architectural form. A group of children playing around the patio; sunshine, raindrops, fallen leaves, air, birds and so on, all these were brought in through the patio. And I want to reproduce this memory in this case.

>> 这里原本只是城市东郊一个垃圾场，我们用了3年植树造林，修复了这个地球小伤疤。建造了10间房，让人们可以在这儿停留，交谈。侥幸在这样一个环境，能让人提及些关于人与自然的话题，那将是我们最期待的事情。人，自然，建筑，应该是怎样一种关系最恰当。记忆中，儿时外公家那个被四周房屋围成的院子，是让我一直无法忘记的建筑形式。一群孩子围着天井追逐嬉戏；阳光，雨水，落叶，空气，鸟儿……透过院子上方的"开口"没有遮挡的进入空间。很想用这个案例来再现这段记忆。

After reaching a consensus with the owner, we launched out this project. The final architectural plane may cause some criticism: what should the owner do when it rains? Does the owner need to walk a long way to the bathroom? And the kitchen is far away. During that period, I weighted repeatedly and asked myself: is the function the paramount thing? Is the habit the necessary thing? In the moment between gaining and losing, I gave up the so-called "absolutely" and decided to choose the smooth and true experience of space.

与业主达成共识后,我们的设计开始了。最终确定并完成的建筑平面,实在会让很多人提出各种常识性的质疑,下雨怎么办?需要走这么长的路去卫生间吗?厨房很远的!在那段日子里,我反复权衡,功能一定是首要吗?习惯永远是必需吗?患得患失间,我放弃了人们口中的那些"绝对",选择了内心对空间细腻和真实的体验。

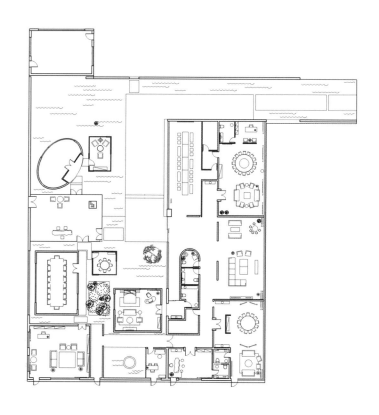

In order to "vanish" the architecture and make it no longer the estrangement between human and nature. We used the further processed surplus soils to accomplish all the facades of the architecture, and the old wood purchased from the second-hand market is the main character of the space. In the morning, when the first hint of sun penetrates the aperture and comes into the room, it shines on the wall made of clay and old wood, the changes and texture made by the shadow is exciting. Almost all the visits can interpret its feeling of Chinese space, which is very delightful as it is something we don't convey deliberately but eager to show. I believe, it is probably the common memory that we have for the "patio".

为了使建筑"消失",让它不再是人与自然的隔阂,我们用深加工处理后的土建余土,完成了建筑所有立面,而那些在旧木市场收购回来的老木头则是室内空间的主角。清晨,当第一缕阳光透过建筑的缝隙进入室内,洒在泥土和旧木制造的墙上,光影让空间产生的变化和质感,让人兴奋不已。几乎所有来过这里的人都能读出中国空间的感觉,这让我非常高兴,因为这的确是一个没有刻意却又很想传递的信息。我想,这可能是大家对"天井"都有的一个共同的记忆吧。

BRITISH & FRENCH STYLE 英法风格
MODERN LUXURIOUS STYLE 现代奢华风格
SIMPLE EUROPEAN STYLE 简欧风格
MODERN CHINESE STYLE 现代中式风格
MEDITERRANEAN STYLE 地中海风格

MEDITERRANEAN STYLE
地中海风格

5

SKY AND SEA
海天一色

设计公司：上海元柏建筑设计事务所
设 计 师：史迪威
项目地点：上海
项目面积：230 平方米
主要材料：地板、木吊顶、装饰画等

DESIGN CONCEPT / 设计理念

>> The beauty of Mediterranean style includes "sea" and "sky", bright colors, white walls as if washed by water, the fragrance of lavender, rose and jasmine, colors of the flower fields, ancient architectures with a long history and the strong national color interweaved by earthy yellow and reddish brown.

>> 地中海风格的美，包括"海"与"天"、明亮的色彩、仿佛被水冲刷过后的白墙，薰衣草、玫瑰、茉莉的香气，路旁奔放的成片花田色彩，历史悠久的古建筑，土黄色与红褐色交织而成的强烈民族性色彩。

The designer presents the permeability of the space through the design of continuous arches and horseshoe-shaped windows, displays the openness through trestle-shaped terrace and open rooms with different functions and expresses the free spiritual connotations of the Mediterranean style through a series of open and permeable architectural decorations. At the same time, the natural materials reflect the life fun of yearning to nature, closing to nature and attaching nature. The sky-blue-based color collocation, the skillful application of natural lights and featured soft decorations with fluent and fancy lines state the romantic moods. The designer adopts loose and comfortable furniture to embody the free, natural, romantic and casual essences of the Mediterranean style.

设计师通过空间设置上的连续的拱门、马蹄形窗等来体现空间的通透，用栈桥状露台、开敞式房间功能分区体现开放性，通过一系列开放性和通透性的建筑装饰语言来表达地中海装修风格的自由精神内涵。同时，通过取材天然的材料方案，来体现向往自然、亲近自然、感受自然的生活情趣。通过以海洋的蔚蓝色为基色调的颜色搭配方案，自然光线的巧妙运用，富有流线及梦幻色彩的线条等软装特点来表述其浪漫情怀。设计师在设计上大量采用宽松、舒适的家具来体现地中海风格自由、自然、浪漫、休闲之精髓。

FRESH BREEZES BLOW GENTLY
清风徐来

设计公司：上海飞视装饰设计工程有限公司
设 计 师：张力
项目面积：450平方米
项目地点：山西太原
主要材料：天然大理石、高级白橡木饰面、高级石材马赛克拼花、高级仿古砖等

DESIGN CONCEPT / 设计理念

>> The style of this project is fresh Mediterranean style. The choices of materials and colors highlight the main tone of the sky blue seacoast and white sand beach in Mediterranean. Whether it is gray tone in the ground, blue tone in the wall or oak open paint in the door packet and stairs, all stress the peaceful, cozy and natural amorous feelings in Mediterranean. As an edge-set house type, the designer focuses on creating the extra windows from the east side through decorative techniques to bring the outside sun lights into interiors and to make the beauty of Mediterranean style present not only in being with sea, blue sky and sand beach, but also in the special feelings from the fusion of sun and nature.

>> 设计风格为清新地中海风格，在选材和配色上，都突显地中海那蔚蓝海岸与白色沙滩的主基调，不管是地面的灰白基调，墙面的蓝色基调，还是门套楼梯等白橡木开放漆的表达手法，都无不重复地突出地中海的祥和安逸的自然风情。作为边套户型，我们通过装饰手法重点打造东侧增加出来的窗户，将室外阳光引入室内，让地中海风格的美不仅表现在与大海、蓝天、沙滩为伍，更有与光与大自然融合的特殊风情。

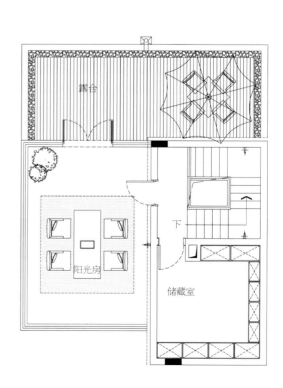

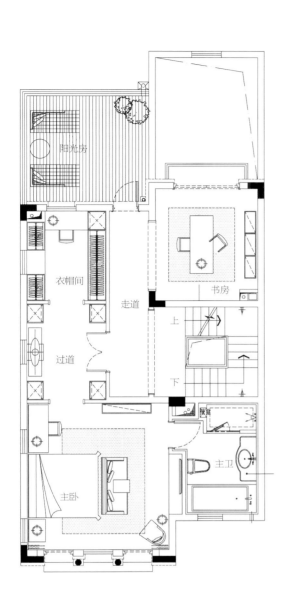

Its color tone is bright, fiery and rich. In addition the prominent national and regional soft adornments make the living atmosphere of the villa full of peaceful, tranquil, free and casual feelings from blue sky and sea. When entering the door after a busy day, the owners can feel a sense of belonging and comfort.

它的基调是明亮、热烈和丰富的色彩，外加民族性地域性特色显著的软装饰品，使得这套别墅的居住环境中充盈着蓝天碧海的安宁静谧、自由自在。从而让业主在辛勤工作一天后，从进门的那一刻开始就能有归属感、舒适感。

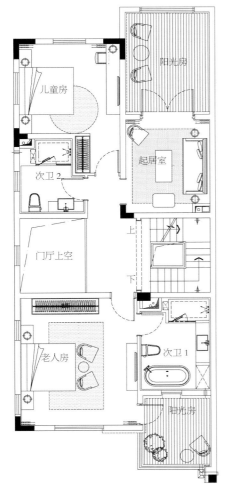

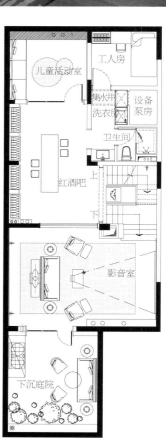

图书在版编目（ＣＩＰ）数据

献计献策：样板房设计新法．Ⅱ．上 / 深圳视界文化传播有限公司编．-- 北京：中国林业出版社，2016.6
ISBN 978-7-5038-8587-7

Ⅰ．①献… Ⅱ．①深… Ⅲ．①住宅－室内装饰设计－图集 Ⅳ．① TU241-64

中国版本图书馆 CIP 数据核字（2016）第 135567 号

编委会成员名单
策划制作：深圳视界文化传播有限公司（www.dvip-sz.com）
总　策　划：万绍东
责任编辑：丁　涵
装帧设计：潘如清
联系电话：0755-82834960

中国林业出版社　·　建筑分社
策　　划：纪　亮
责任编辑：纪　亮　王思源

出版：中国林业出版社
（100009 北京西城区德内大街刘海胡同 7 号）
http://lycb.forestry.gov.cn/
电话：（010）8314 3518
发行：中国林业出版社
印刷：深圳市汇亿丰印刷科技有限公司
版次：2016 年 7 月第 1 版
印次：2016 年 7 月第 1 次
开本：230mm×300mm，1/16
印张：20
字数：150 千字
定价：398.00 元（USD 79.00）